Daniel Wandera C. Naku

TOWARDS SUSTAINABLE URBANISM

Daniel Wandera C. Naku

TOWARDS SUSTAINABLE URBANISM

A NEW URBAN AGENDA FOR RAPIDLY URBANIZING AFRICA

LAP LAMBERT Academic Publishing

Cover image: www.ingimage.com

Publisher:
LAP LAMBERT Academic Publishing
is a trademark of
Dodo Books Indian Ocean Ltd. and OmniScriptum S.R.L publishing group

120 High Road, East Finchley, London, N2 9ED, United Kingdom
Str. Armeneasca 28/1, office 1, Chisinau MD-2012, Republic of Moldova, Europe
Managing Directors: Ieva Konstantinova, Victoria Ursu
info@omniscriptum.com

Printed at: see last page
ISBN: 978-620-8-44116-6

TOWARDS SUSTAINABLE URBANISM

A NEW URBAN AGENDA FOR RAPIDLY URBANIZING AFRICA

AUTHORED BY:

DANIEL WANDERA CLIEF NAKU (PhD)

FORWARD

There is no doubt that African cities are rapidly urbanizing and consequently impacting the rate of urbanism in these cities. In fact, during the last two decades, Africa has experienced the highest urban growth rate at 3.5% per year and this growth rate is expected to hold into 2050.This situation makes rapid urbanization to be one of the most pressing challenges that need to be handled through effective policies and clear practical affirmative actions.

Arguably, urbanism which mainly involves the observed interaction between the urban population and the built environment is indeed becoming complicated phenomenon to be contained in most African cities.

Hence, while the urban planning profession is expected to ensure that urban areas are well planned to provide for better living standards for the urban populations, there is no doubt that under the current ongoing rapid urbanization, most African Cities are under a lot of pressure to the point of collapsing.

In this book, it is noted that as urbanization increases without profound and strategically initiated plans in place to moderate it, the principles of urbanism that include walkability, safety, connectivity and livability among others are constantly being compromised and eroded.

In the long run, this is pushing most of the cities to experience urbanism which is not sustainable in nature since the observed urbanism in most cities does not align itself with the principles of urbanism from the urban planning point of view.

For that matter and on account of the current situation facing many of our African cities, it is the desire of the author of this book under the theme ***'Towards Sustainable urbanism: A New Urban Agenda for Rapidly Urbanizing Africa'*** to shed light on the subject matter. Precisely, the author's intention is to shed light that will influence future urban development policies that are in favor of sustainable urbanism in our cities and not simply the urbanism being observed as of today. It should be noted that Cities are where development challenges and solutions meet and for that matter, this book is set to drive towards this direction.

It is therefore my hope that readers of this book will find it to be a reading material especially among the urban planners, policy makers and students. It is anticipated that the book will provide them with the basis for some of the decisions and initiatives that they ought to undertake in the interest of making urban cities livable and more precisely enhance sustainable urbanism in our cities.

ABOUT THE BOOK

This book views transition to sustainable urbanism as a new urban agenda for rapidly urbanizing Africa. In a sense, the book kick-starts the process of determining pathways to sustainable urbanism for Africa's rapidly urbanizing cities.

In addition, book also succinctly seeks to inform the reader on the various perspectives concerning urbanism, the state of urbanism in Africa, the concept of sustainable urbanism as a new urban agenda for Africa's rapidly urbanizing cities and the possible ways for institutionalizing sustainable urbanism in Africa's rapidly urbanizing cities.

The book's format as reflected in each part, is such that it will facilitate teaching and learning. It is therefore hoped that, it will become a major reference material for teaching and learning using our immediate urban scenarios as the laboratory. It is therefore anticipated by the author that this book will be used intensively by academicians, Planners, policy makers, researchers and students in comprehending the book's subject matter for the benefit of promoting sustainable urbanism.

ABOUT THE AUTHOR

Dr. Daniel Wandera Clief Naku (PhD) is a lecturer in the Department of Geography at the St John's University of Tanzania, Dodoma. He holds a First Class Hons Degree in BSc Urban and Regional Planning from the University of Dar Es Salaam (UDSM), Tanzania; a Master of Science Degree in Development Planning and Management from TU-Dortmund, Germany and Kwame Nkrumah University of Science and Technology (KNUST); a Master Degree in Development Economics from the Institute of Rural Development Planning (IRDP), Tanzania as well as a PhD in Policy Analysis and Programme Management from Ardhi University (ARU), Tanzania. He has over 15 years of working experience which involve university teaching, research, consulting and providing professional training. His research interests include Urban and Regional Development Policy, Local Economic Development, Policy Studies, Community Engagement/Participation as well as socio-economic impact and monitoring studies. He also offers professional training in aspects of organization leadership, monitoring and evaluation of development programs and projects, grant proposal writing, project proposal and concept note writing as well as strategic planning. His publications include four books titled; *Project Analysis and Evaluation for Development Practitioners and Academicians: A Guide to Project Analysis and Evaluation of Development Projects; Going Strategic at the Community Level: An insight into the how's of strategic planning at the community level; Project Concept Note and Proposal Writing Guide: The Art of Writing Project Concept Notes and Proposals for Development Projects as well as Extending the Frontiers of Knowledge through Research and Publications. A Research Repository Book.*

COVERAGE AND STRUCTURE

The book consists of six parts that seek to enable the reader to comprehend the subject matter on Urbanism and Sustainable Urbanism. Part One of the book presents the Introduction. The second part covers Perspectives on Urbanism. Part Three discusses the State of Urbanism in Africa and Urbanization as its major driving force. Part four provides insights on Sustainable Urbanism as a new Urban Agenda for Africa's rapidly urbanizing cities. Part five discusses Institutionalizing Sustainable Urbanism in Africa's rapidly urbanizing cities and finally part six winds up the book with Concluding Remarks.

ACKNOWLEDGEMENT & DEDICATION

In a very special way, I want to acknowledge God's favor for the knowledge and wisdom he has bestowed upon me. James 1:5 states that; *if any of you lacks wisdom, let him ask God, who gives generously to all without reproach, and it will be given to him.*

I was motivated to write this book by my students who are required to study a course module on urban systems at the St John's University of Tanzania. As such, I wish to acknowledge these students for their motivation and hope that this book will play a foundational role in their education endeavors and quest for knowledge on urban studies.

I am also indebted to Professor Einhard Schmdit –Kallert of TU-Dortmund University in Germany and Professor. Tiberio Mdendemi from the Institute of Rural Development Planning in Tanzania for their review of the book and rendered constructive comments.

Finally, this book is dedicated to my late father Christopher Emmanuel Wandera and late mother Phelomera Anyango Wandera. May you both continue to Rest in Eternal Peace.

WISDOM WORDS

Knowledge is a process of pilling up facts, wisdom lies in the simplification **Alexander Graham Bell**

We are what we repeatedly do, Excellence, then, is not an act, but a habit **Aristotle**

We faced hard days and our share of failure, but I learned then that no matter how great a challenge may be or how difficult the circumstance, change is always possible if you are willing to work for it, and fight for it, and, above all believe in it **Barack Obama**

For if you know these things about a man you can touch him at the core of his being **Bill Bernbach**

Coming together is a beginning, keeping together is a progress, and working together is a success **Henry Ford**

If I have no one to read, I have no reason to write.
Inioluwa Ezekiel

CONTENTS

PART ONE

INTRODUCTION

General Introduction

I wish to begin the introduction of this book with an English saying that 'Not all that glitter is gold'; and so is the case with the ongoing urbanization process in Africa. Literature as revealed by Konrad (2020) shows that the African continent is experiencing rapid urbanization and this has become a point of discussion among scholars, planners and policy makers. For instance, it is reported by Ramanatha (2019) that during the last two decades, Africa's urbanization rate has been at 3.5% per year and this growth rate is expected to hold into 2050.

Unfortunately, as stressed by the African Development Bank Group (AFDB, 2012), the ongoing urbanization in Africa is largely manifested through rising slum establishments, increasing poverty and inequality. In addition, there are large variations in the patterns of urbanization across the African continent. For instance, according to AFDB (2012), North Africa has a higher proportion of urban population at 47.8% compared to Sub Saharan Africa (SSA) whose proportion of urban population stands at 32.8%.

A major difference between the ongoing urbanization in North Africa and Sub Saharan Africa as observed by the African Development Bank Group (AFDB, 2012) is that the urbanization in North Africa is characterized by fewer slums which is attributed to implementation of effective urban development plans and policies. In contrast, the urbanization in Sub Saharan Africa as noted by UN-Habitat (2020) is dominated by highest proportion of slum dwellers (65%), insufficient basic infrastructure and only 20% of SSA's population has access to electricity. In addition, 60% of African citizens according to UN-Habitat (2020) live in places where water supplies and sanitation are inadequate. To a large extent, the manifested urban situation in SSA is attributed to limited investment in infrastructure and the failure to implement effective urban development plans and policies capable of steering urbanization in the right direction.

Amidst such urbanization challenges, realizing sustainable urbanism as defined under the principles of urbanism is becoming a far-fetched dream in most African Cities and yet still something that can be achieved if governments opt to pursue key reforms that focus on upgrading informal settlements, mobilizing urban financing from local and foreign investors, improving human capital and diversification of urban economic activities among others.

On the other hand, the time is right for planners, policy makers and scholars to explore what works and what does not work in the African urbanization process. By so doing, managing urbanization in the context of Africa will be governed by how Africa defines urbanization and the standards by which such urbanization is measured.

It should also be noted that while there is a general view and desire to copy and replicate what is happening elsewhere and especially from Europe like in the UK and Asia like in China, dealing with urbanization issues on the African continent will need more locally derived solutions than imported solutions.

In general terms, it is presumed in this book that once urbanization is contained, realizing sustainable urbanism may be achieved. At this point, one will expect cities to livable, internally well connected, walkable, clean, environmentally aesthetic and above all safe for everybody. Indeed, efforts in this direction are yet to be crystalized but the need cannot be underestimated for our African cities.

PART TWO
PERSPECTIVES ON URBANISM

Overview

This part of the book draws the reader's attention to various perspectives on urbanism. Precisely, the intention is to enable the reader to grasp and appreciate what urbanism is about based on the various competing views.

Perspectives on Urbanism

Generally, urbanism is a concept that is sometimes found to be difficult to define. In fact, some scholars who have tried to define it, have found their definitions fall short of something in terms of clarity or overgeneralization. In the context of this book, we will try to view some of the perspectives that have been put forward in an attempt to bring to light what urbanism is about.

To begin with, literature shows that, the most consistent definition of urbanism according to Raptis (2017) comes from Louis Wirth a sociologist in 1938. According to Louis Wirth, urbanism is described as a characteristic mode of life that may be approached empirically from three interrelated perspectives.

First as a physical structure comprising a population base, a technology, and an ecological order. Second as a system of social organization involving a characteristic social structure, a series of social institutions, and a typical pattern of social relationships and lastly as a set of attitudes and ideas, and a constellation of personalities engaging in typical forms of collective behavior and subject to characteristic mechanisms of social control.

From Louis's definition, we note that, although Louis Wirth came up with this definition in 1938 and envisioned the urban environment according to those times, his definition explains very well which perspectives must be considered when discussing urbanism. In this regard, urbanism needs to be seen as an approach to life in cities, which is the result of social concentration and masses of people gathering to create societies.

It explores the behavior of social concentration and the gathering of masses of people. Hence, since this behavior has existed for as long as humankind, consequently urbanism has been a topic of thought and practice from the very beginning. To this regard, as humankind evolves and so does urbanism. This can be manifested in the multitude of urban environments in the planet, both in a historical and a geographical context.

As a physical structure, Louis Wirth described one of its perspectives as being always in development. A multitude of different urban environments have existed throughout mankind's existence on this planet and it constantly evolves as it takes advantage of the newest technology.

As a system of social organization, urbanism approaches the effects this has on urban life. For example, the analysis of the different effects that capitalistic and communistic societies have on the environment. Similarly, social patterns can change the nature of a place. For instance, some places have a high concentration of social dynamics and as such they constitute very important urban spaces.

Similarly, urbanism as a set of attitudes and ideas, describes what Louis Wirth observed in his time. According to Louis, urban areas tend to demonstrate collective social behavior as opposed to rural areas. In the aspect of collective social behavior, Louis stresses that cities tend to act as hubs of such behaviors. This is because in places with larger masses of people, there is higher probability of finding common social ideas and creating social norms.

In addition, cities tend to create dynamic social existence, between different and counter acting ideas which in turn shape the places around them. As a result, urbanism is seen by many as a special way of living, although types of urbanism have existed for centuries. It is dynamic, always to some degree extending its influence outward. However, what needs to be stressed is that modern urbanism is even more dynamic and more outreaching in character.

Urbanism is also sought to involve the understanding and addressing of various aspects of urban life, such as social, economic, environmental, and spatial elements, to create sustainable, functional, and livable urban environments. The concept of urbanism recognizes that cities are complex systems with interconnected elements and aims to improve the quality of urban life by shaping and managing these elements.

It therefore considers the diverse needs and aspirations of urban dwellers and strives to create inclusive, equitable, and vibrant urban spaces. Similarly, urbanism seeks to create cities that are socially inclusive, economically prosperous, environmentally sustainable, and aesthetically pleasing. To this regard, urbanism tends to combine various disciplines and perspectives to address the complex challenges and opportunities associated with urbanization, ultimately aiming to improve the quality of life for urban resident.

Besides Louis's views on what urbanism might mean, other scholars as pointed out by Moroni (2016) also came up with different views on the matter. For instance, according to Jane Jacobs an urbanist, urbanism is seen as looking at cities as living organisms, and it stresses the importance of diversity, flexibility, and adaptability in urban design. The French philosopher and sociologist Henri Lefebvre have viewed urbanism as a social and cultural phenomenon that arose with industrialization, and which produced new forms of space, time, and social relationships.

On the other hand, a U.S. geographer Edward Soja as pointed out by Soja (2009) described urbanism as a complex network of social, economic, and political forces that shape the spatial organization and distribution of power within cities. A planning scholar Kevin Lynch as cited by Vlahov et al (2010) looks at urbanism as the art and science of producing good urban form which balances a range of factors such as aesthetics, functionality, and social interaction but Blokland-Potter, (2008), viewed urbanism as is the way of life.

To this regard, Haferkamp and Smelser (1992) stress that urbanism reflects organization of societies in terms of complex division of labor, high level of technology, high mobility, and interdependence of its members in fulfilling economic functions and impersonality in social reality. Siravo, (2019) summed up the above perspectives on urbanism into six perspectives as explained below.

Social perspective: This perspective emphasizes the importance of creating cities that are socially inclusive and equitable, where all residents have access to basic services, affordable housing, and quality public spaces. It also focuses on fostering social cohesion and community-building. Three Major Perspectives in Sociology Sociologists analyze social phenomena at different levels and from different perspectives.

Sociologists today employ three primary theoretical perspectives: the symbolic interactionist perspective, the functionalist perspective, and the conflict perspective. These perspectives offer sociologists theoretical paradigms for explaining how society influences people, and vice versa. Each perspective uniquely conceptualizes society, social forces, and human behavior. But also sociology have their perspective on urbanism which is the functionalist perspective.

According to the functionalist perspective, also called functionalism, each aspect of society is interdependent and contributes to society's functioning as a whole. Functionalists believe that society is held together by social consensus, or cohesion. Not necessarily the case for latent functions, which often demand a sociological approach to be revealed.

A sociological approach in functionalism is the consideration of the relationship between the functions of smaller parts and the functions of the whole. Functionalism has received criticism for neglecting the negative functions of an event such as divorce. On the other hand, critics of the conflict perspective which originated primarily out of Karl Marx's writings on class struggles, presents society in a different light than do the functionalist and symbolic interactionist perspectives.

Environmental perspective: This perspective highlights the need to design cities in a way that minimizes their impact on the natural environment, promotes sustainability, and enhances the quality of life of urban dwellers.

Economic perspective: This perspective emphasizes the importance of creating cities that are economically vibrant and competitive, where businesses can thrive and generate jobs and wealth. Basics of Economic Principles Economics is the field of study that examines how resources are used to produce goods and services and how purchasing decisions are made by consumers based on scarcity.

Understanding basic economic principles such as scarcity, supply and demand, marginal costs, marginal benefits, and incentives are important to economic decision-making for producers and consumers. A consumer (or buyer) is an individual, company, organization, governmental body, or other entity that acquires or purchases goods or services in the economy. Economic producers are the individuals or organizations that use factors of production to manufacture or provide goods or services for consumers to purchase. Scarcity refers to the economic principle that the world has limited resources, and resources must be allocated in an organized fashion.

The principle of supply and demand explains how prices of resources are dependent upon supply and consumer demand for resources. Consumers will make purchasing decisions based on costs, benefits, and incentives. Economic theory states that consumers use economic decision-making principles such as trade-offs, giving something up to get something, marginal cost analysis, and consumer response to incentives when making purchasing decisions.

Marginal benefit is the expected satisfaction that a customer expects to receive as a result of purchasing a good or services. A consumer choosing between a name brand or value brand box of cereal can look at the price per ounce that is expressed on the display tag. This cost per ounce helps the consumer decide which brand is the cheapest per unit and weigh the price into the purchasing decision.

Cultural perspective: This perspective recognizes the importance of preserving and promoting cultural heritage and diversity, as well as creating cities that foster creativity, innovation, and artistic expression. Cultural studies are the critical examination of popular culture, which is understood as any culture with widespread appeal, from the arts and entertainment to advertising and marketing.

Cultural studies scholars investigate how cultural messages are created, distributed, and consumed and their effects on society. The basic principles of cultural studies are that culture is created by humans and shapes, and is shaped by, society. Additionally, cultural studies believe that culture should be studied to understand both organization and the individual.

Cultural studies have four critical methods for understanding culture: textual analysis, historical analysis, ethnography, and discourse analysis. Each method provides a different perspective on how culture can be understood.

Additionally, cultural studies adopt a critical perspective that examines power relations within societies; this critical perspective allows for examining how various forms of oppression are reproduced through cultural texts and practices.

Political perspective: This perspective emphasizes the role of political institutions and governance structures in shaping urban development and ensuring that urban policies and decisions are responsive to the needs and aspirations of all residents. In this perspective, five principles serve as the foundations of the political perspectives.

The Rationality Principle holds that all political behavior has a purpose and that people tend to be goal-oriented in their political activities as they make instrumental choices about how to act. The Institution Principle recognizes that political institutions provide incentives for political behavior.

The Collective-Action Principle refers to the idea that, although all politics is collective action, getting people to act in concert is difficult. Thus, political action often involves both formal and informal bargaining, along with other efforts to overcome the obstacles to collective action.

Formal and informal bargaining relationships are struck in politics to provide for collective decision making, and organizational efforts to overcome the tendencies of individuals to "free ride" on the labors of others are necessary to overcome the collective-action problems endemic in producing public goods.

The Policy Principle holds that political outcomes are the products of individual preferences and institutional procedures. Combining lessons from the Rationality and Institution principles, we see that individual political actors' personal, electoral, and institutional ambitions are filtered through, and in many ways shaped by, institutional arrangements in politics; and that policy.

Historical perspective: This perspective recognizes the importance of understanding the historical evolution of cities and the ways in which past decisions and policies have shaped their current form and function. It also emphasizes the need to preserve and protect historic urban landmarks and sites.

It offers surprising alternatives to the taken-for-granted, conventional wisdom, and opens a wider perspective from which to evaluate our present preoccupations. Hence, taking historical perspective means understanding the social, cultural, intellectual, and emotional settings that shaped people's lives and actions in the past.

At any one point, different historical actors may have acted on the basis of conflicting beliefs and ideologies, so understanding diverse perspectives is also a key to historical perspective-taking.

Though it is sometimes called "historical empathy," historical perspective is very different from the common-sense notion of identification with another person. Indeed, taking historical perspective demands comprehension of the vast differences between us in the present and those in the pasts.

Hence, some of the definitions on what urbanism could be as summed up from literature by Raptis (2017) are as presented in the Table 1 below;

Table 1: Some of the Definitions of Urbanism

S/NO	Source	Definitions
1	Webmasters New World College Dictionary	• The character of life in the cities; urban life, organization, problems, etc. • Movement of the population to, or concentration of the population in, the cities
2	Old Wikipedia definition	• The study of cities - their geographic, economic, political, social and cultural environment, and the imprint of all these forces on the built environment.
3	The American Heritage Dictionary	• The culture or way of life of city dwellers. • Urbanization. • The theory of city planning.
4	New Wikipedia definition	• The study of the characteristic ways of interaction of inhabitants of urban areas with the built environment. It is a direct component of disciplines such as urban planning and urban sociology

Generally, based on the above explanations, urbanism is indeed a broad and complex concept. Its meaning cannot be attached to a single factor and neither can we assume it to be a static phenomenon. Rather, it is a dynamic phenomenon which changes with time and influenced by the political, social, cultural, environmental and economic circumstances in a given urban area.

In addition, while the understanding of urbanism as explained by the various scholars may seem to vary, it is evident that all scholars appear to see urbanism as a driving force that in one way or another is a manifestation of the urban life. Such manifestation could be in the political, social, economic and cultural aspects of that particular urban area.

Hence, in concluding this section, it is important to note that this book acknowledges the fact that, there are several perspectives on what urbanism could be about. However, major emphasis has been put on the most common perspectives that may enable a reader to grasp the concept of urbanism in its totality.

PART THREE

STATE OF URBANISM IN AFRICA AND URBANIZATION AS ITS MAJOR DRIVING FORCE

Overview

This third part of the book seeks to draw the reader's attention to the state of urbanism in Africa and the driving forces for such urbanism. Precisely, the writer's desire is to enable the reader to obtain a clear picture on the aspect of urbanism in Africa and the factors that are contributing to urbanism in African cities.

Urbanism in Africa

To understand urbanism in Africa i.e. a way of life, we will need to explore this aspect from three dimensions that include size, population and dominating functions in African cities.

From the size dimension, African urbanism is largely characterized by small- and medium-sized cities as revealed by Dodman et al, 2017. In fact, according to updated data from global geografia for the period 1998-2023, the largest city in Africa is Johannesburg from South Africa covering an area of 4,040Km^2 while the smallest city is Bunia from the Democratic Republic of Congo covering 60Km^2. This growth in size is largely influenced by several factors but the most obvious being natural population increase among city residents and migration to cities from rural areas.

Despite of this increase, most of the cities are below 500 Km^2 which implies that majority of African cities are either small or medium in size. A situation that could be attributed to both historical and political definition of boundaries of settlements.

In the interest of this book, Figure 1 below shows only those cities that are considered to have a coverage of $500Km^2$ and above. Categorically, these cities fall in the category of medium and large which also implies that they play a vital role in influencing economies of their countries and that of the African continent as a whole.

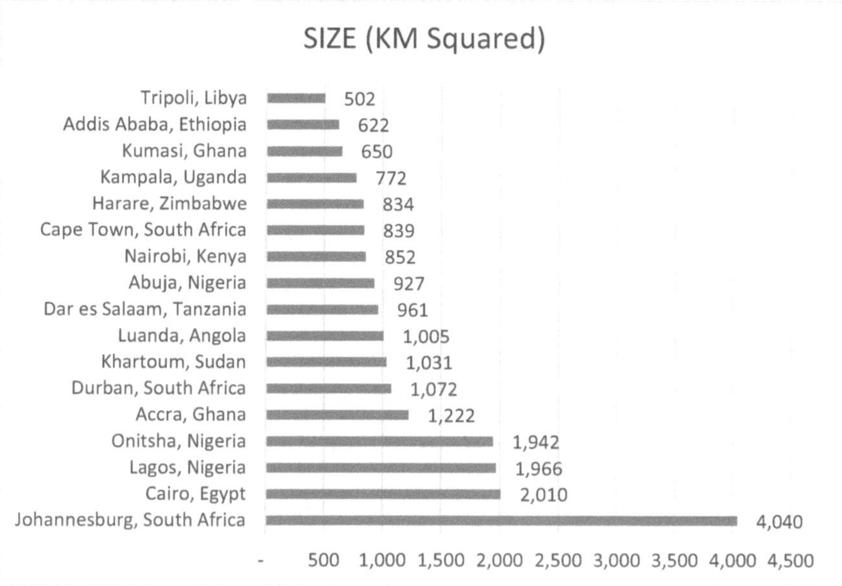

Figure 1: Sampled Cities by Size

With most of Africa's cities having a coverage of below 500 square meters, the economies of these cities is equally small. This limits such cities from capitalizing on the greater scale of economies that is normally enjoyed by larger cities (Dodman et al 2017).

In terms of population as pointed out by the African Development Bank (2012), African cities are experiencing the highest rapid population growth at the rate of 3.5% per year. This rate of growth is expected to hold until 2050 and the projections also indicate that by 2025, some African cities will account for up to 85% of the population. Such movements mainly involve the moving of many people from the rural settings to urban areas which sets in place a number of challenges.

For instance, most of the migrants from rural areas are uneducated/unskilled. As a result, they end up in the informal sector which accounts for 93% of all new jobs and 61% of urban employment in Africa. In addition, since the incomes from the informal sector are by their very nature low and intermittent, most migrants naturally seek for shelters or become tenants of slum landlords. This has led many African cities to have to deal not only with slum proliferation but also with increasing insecurity and crime. To illustrate the population phenomenon in African cities, Figure 2 presents some of the most populated African cities

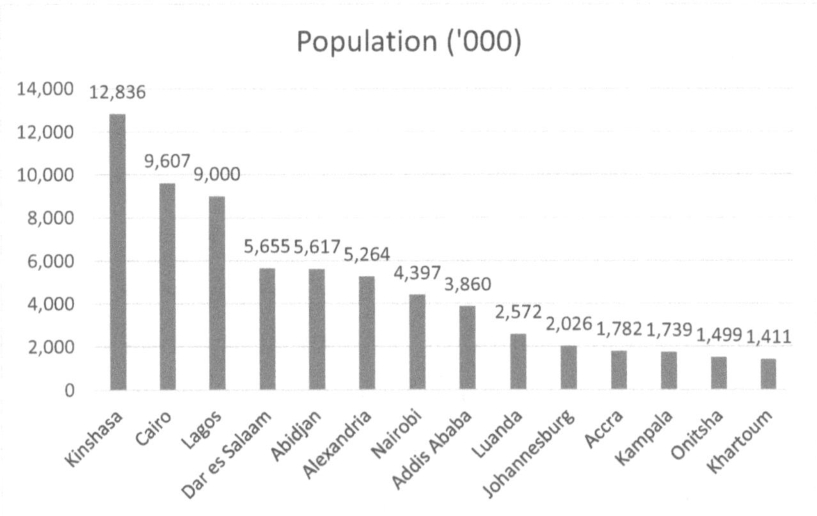

Figure 2: Most Populated Cities on the African Continent

While population growth may be a necessity for the flourishing of demand and supply in any given city, a population that does not fit well in the economics of a city may be a burden rather than an opportunity. Hence, considering that a huge portion of the urban population in most African cities is from the rural settings with no skills that meet the needs of urban economies in terms of jobs, most urban areas end up struggling with accommodating largely unskilled population. Ideally, this kind of population is not a resource base for urban areas to thrive. As a result, economic functionality of most African cities are limited to secondary based economic activities which largely operate in the informal sector.

Figure 3: Some of the functional activities in Dodoma City, Tanzania

Indeed, there is a strong link between population and economic functionality of cities. We have seen that the population in a given city based on possessed skills can determine the kind of economic activities carried out in that city. This being the case, we now move on to examine the third dimension of urbanism in Africa. As previously stated, urbanism in African cities can also be explained by observing the dominating functions of a particular city. It is these functions that manifest what people in that city do and thus spend most of their time doing to earn a living.

On the functional dimension, it should be stressed that data on economies of African cities is limited in comparison with most other parts of the world. As a result, these cities are often perceived as overcrowded, congested and unproductive. Nevertheless, it is evident that many African cities are not well planned, lack infrastructure and provide insufficient public services, compared to cities in other parts of the world. They face increasing levels of pollution and are threatened by climate change. The high costs of doing business reduce the competitiveness of their economies, and African cities are not experiencing the rapid transformation seen in cities in other emerging economies, such as China.

In fact, most African cities are dominated by low productivity, informal enterprises and mostly engaged in the service sector (Danquah & Kunal, 2023). It is in this sector that large segments of Africa's urban population work with low pay and often self-employed. In addition, in these cities, prospects of a thriving manufacturing industry are limited and even the share of employment in the manufacturing sector is equally low. For instance, in the three original East African States of Uganda, Kenya and Tanzania, data shows that a huge share of employment when compared between the service sector and the industry sector lies in the service sector.

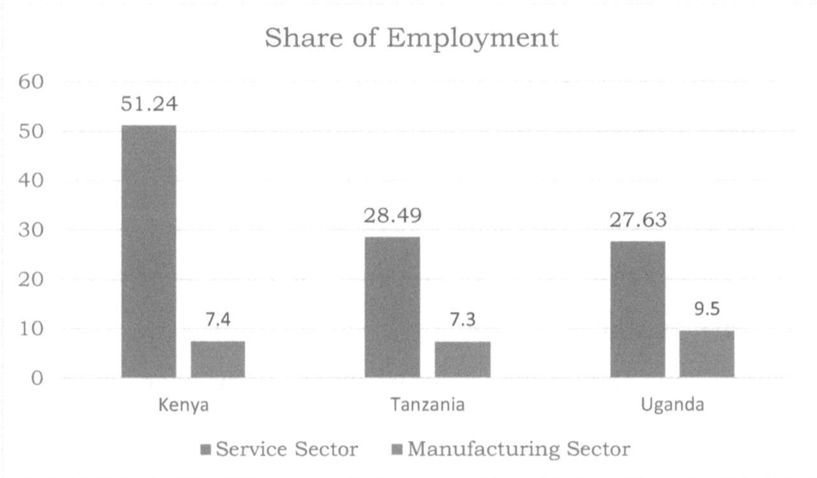

Figure 4: Share of Employment in Sampled African Countries

From the above illustration, it can be noted that the service sector is such a dominating sector in the examined countries. However, in its dominance, majority of the employment in this sector is found in the informal sector as opposed to the formal sector. This explains the low economic productivity of most African cities and their inability to enjoy scale of economies when compared to their counterpart cities found in Europe.

Taking into consideration the nature of economic activities carried out by the majority of urban residents, from the functional point of view, it can be arguably stated that urbanism as a way of life in most African cities is largely manifested in the service sector with an informal structural set up. This being the case, then we are faced with a key question on what exactly drives this kind of urbanism in African cities.

Key Features of African Urbanism

In the interest of this book, the author finds it of necessity to discuss some of the key features of the ongoing urbanism on the African continent. In fact, a quick assessment of these features show that the ongoing urbanism on the continent has negative manifestation. This manifestation is demonstrated by the following features;

Small and Medium Sized Cities

As previously expressed, African urbanism is largely characterized by small- and medium-sized cities as revealed by Dodman et al, 2017. In fact, according to updated data from global geografia for the period 1998-2023, the largest city in Africa is Johannesburg from South Africa covering an area of 4,040Km^2 while the smallest city is Bunia from the Democratic Republic of Congo covering 60Km^2.

While in a Local and national governments meeting convened at the 9th Africities gathering in Kisumu from 17 to 21 May 2022 concluded that mid-sized cities spread the benefits of access to cities to a large and dispersed rural population in Africa, it is also important to note that such cities have become a burden to administer and certain groups of people especially the youth and women are not fully integrated into plans of these cities. Considering that these cities serve as Intermediate cities between the big cities and rural areas, they should focus on the development of women and young people as well as on their infrastructure and modernization.

Informal Settlements

This is another important feature of the ongoing urbanism on the African continent. According to the African Cities Research Consortium, in most African cities, more than half of the residents live in informal settlements. Statistically, Ono and Tetsuo estimate that 30%-70% of the African urban population live in informal settlements, which is considered to be a response to the failure of government controlled urban planning systems to meet the housing needs of urban populations. Below is a sample of an informal settlement in Kenya.

Figure 5: Kibera informal settlement in Nairobi, Kenya. Photo: Boris Golovnev/Shutterstock

It should be noted that the emergence of informal settlements in African cities can be traced back to the colonial era, when colonial urban planning divided the urban space into two distinct zones: a "European" space that enjoyed a high level of urban infrastructure and services and an "indigenous" space that had marginal services. Indigenous space, or informal settlements, became a prevailing mode of urban development across Africa in the wave of rapid urbanization after World War II.

In the 1960s, as urbanization was addressed in the theory of modernization as revealed by Ono and Kidokoro (2020), informal settlements were seen as an obstacle to building modern cities. The dominant idea was that informal settlements should be eradicated from urban areas by government-led capital-intensive housing provisions, which ended in failure. In the early 1970s, a new paradigm and urban policy were provided by John Turner, a British architect who argued that housing was best provided and managed by those who were to dwell in it rather than being centrally administered by the government (Ono and Kidokoro, 2020).

John Turner showed that self-designed settlements by local groups worked better as people were experts on their own situations and should be given the "freedom to build." Within this framework, the government and the private sector act as enablers.

Nevertheless, in the interest of this book, it is strongly suggested that key reform policies need to be put in place to respond to these unfortunate situations from the past.

Inadequate social services

The ongoing urbanism in most cities on the African continent is also characterized by inadequate social services in these cities. According to the World Bank report of 2015, revealed that 90 percent of Africans live in informal housing, where living conditions are often substandard, unsafe and without basic services like water, electricity and sanitation. As a result, as revealed in a recent report of the International Federation of Red Cross and Red Crescent Societies, the lack of adequate social services relating to sanitation, potable water and electricity, in addition to substandard housing and overcrowding, aggravates the spread of diseases and avoidable deaths.

Hence, in the context of this book, it is therefore evident that addressing the issue of inadequate social services in African societies should be considered as a priority if the ongoing urbanism is to be of relevance to the city residents. This will require targeted interventions in the informal market that can bring rapid improvements in the quality of social services and housing in the cities.

Poor and inadequate infrastructure

The fact that Africa's infrastructure is central to its future is well understood by African policymakers. Unfortunately, ensuring that there is good and sufficient infrastructure in most cities on the African continent remains a dream. It should be noted that Africa's infrastructure stocks inherited from the colonial powers supported reasonably strong economic growth from the early 1960s until the 1970s oil shocks. Between then and the mid-1990s, a long economic slowdown, combined with growing interest in regional trade and other economic agreements catalyzed change in Africa's economic structure.

The fast, steady and continuous growth in household demand for infrastructure stemming from Africa's high population growth and the fast increase in the urbanization rates of these populations contributed to fuel the demand independently of all the changes on the production structure of the economy. These changes led to a growing mismatch between the demand and the supply for infrastructure in the region.

By the end of the 1990s, the gap had grown so significantly that some estimates suggest that to meet the Millennium Development Goals (MDGs), the average annual infrastructure expenditures in Africa (the sum of investment and maintenance expenditures) need to be around 9% of GDP between 2005 to 2015.

According to the 2005 Global Monitoring Report, if Africa is to meet its infrastructure development needs, the region needs average growth rates of over 7% for the next 10 years or so. In addition, according to the Africa Economic Outlook report (2018), New estimates by the

African Development Bank suggest that the continent's infrastructure needs amount to $130–170 billion a year, with a financing gap in the range $68–$108 billion.

Generally, based on the above revelations, it is evident that developing key infrastructure on the African continent is still such a huge challenge. As a result, most African cities have become victims of this challenge which has rendered infrastructure to be poor and inadequate in these cities. A Sample of the road infrastructure in Kampala-Uganda.

Figure 6: Poor Road Infrastructure in Kampala City, Uganda

Unplanned and un coordinated spatial expansion

Unplanned and uncoordinated spatial expansion also stands out as a characteristic feature of the urbanism in Africa cities. According to Burak et al (2017), much of urban expansion in Africa is characterized by unplanned and unregulated growth, exacerbated by the legacy of colonialism, structural adjustment, and neo liberalism that spawned weak urban planning institutions. Such expansion lead to ecological imbalances in the cities thus making them not conducive for human settlement.

For instance, as pointed out by Msuya et al (2020), Dar es Salaam is the second fastest growing city in the world and the fifth largest city in Africa. However, over the past 20 years, the city has experienced substantial increases in both built-up areas and population size. The city is becoming densely populated. While the population is deconcentrating in and around the commercial business district (CBD), peripheral wards are seeing high population growth linked to both migration and a population shift from the CBD and its vicinity.

Similarly, Kampala City like other African cities is one of the cities facing unplanned and uncoordinated development. According to a report by Wandera (2014), the land use change in Kampala was changing in unplanned manner leading to incompatible land use activities. Wetlands and swamps in the area had for example been increasingly settled, both informally by the urban poor and formerly by industrial and commercial actors. Below is a sample of the unplanned settlement at Mji Mpya in Dar es Salaam, Tanzania.

Figure 7: Unplanned Settlement at Mji Mpya in Dar es Salaam, Tanzania.

This kind of development undermines efforts to achieve ecological balance in these cities and thus rendering the cities' environments not conducive for human settlement.

Economically weak cities

Africa's economic future lies in its cities. As a result, governments ought to ensure that national development strategies and plans center on the urban economy. In fact, each year, a large number of young people move into cities with hopes of findings jobs and living a better life. Unfortunately, experience shows that most of the cities on the continent have extremely weak economies that cannot accommodate the pressure from the in-coming young people.

In addition, the performance of a city's economy is determined by its productivity which implies increase in quantity and quality of a city's outputs. With regard to this matter, again, evidence shows that productivity in cities of most African countries is both low in terms of quantity and quality.

This situation is highly attributed to the small-medium level of investments that are carried out in cities. Consequently, the economies of most cities on the African continent are highly dominated by the informal sector which is also of small-medium character. Hence, under such circumstances, it is very difficult for economies of these cities to be strong and compete favorably with strong cities in highly developed economies.

Under such circumstances, local development policies that seek to upgrade existing sectors and activities are more likely to succeed in addressing this challenge than local economic development policies that aim at attracting entirely new economic sectors. In addition, concerted policy packages that combine coordinated interventions can be sought as opposed to isolated measures. This is because, such interventions tend to involve multiple policy sectors as well as government and non-governmental actors.

Finally, it should also be noted that different places have different competitive advantages that can be the basis for economic development. As a result, it will be of great relevance to identify and utilize these competitive advantages as key features for strengthening economies of cities on the African continent amidst the on-going urbanism.

High rural-urban migration

Migration phenomenon has been critical to economic progress, education, and mobility. This global phenomenon has been reshaping cultures for ages in countries all over the world. While migration has affected urban growth and development, it also brings challenges to both rural and urban areas.

Rural-to-urban migration refers to people moving from rural to urban areas in quest of better opportunities, higher living standards, and access to critical services. As such movements occur, the gap between rural and urban communities also grows. Young people seeking opportunity and change are continuously moving from their rural areas to the big cities, where the promise of expanded professional options, greater financial independence, and a new lifestyle typically outweigh the disadvantages.

As a result, the surge in population caused by these movements often leads to several challenges as detailed below.

Urban overcrowding

As rural migrants flock to urban centers, cities often experience population surges that lead to overcrowding. This puts immense strain on existing infrastructure, housing, transportation networks, and public services, resulting in challenges related to sanitation, waste management, and traffic congestion.

Housing Challenges

The influx of rural migrants in search of better opportunities drives up the demand for housing in urban areas. As a result, property prices escalate, and affordable housing becomes scarce, leaving many low-income families and migrants vulnerable to living in informal settlements and slums. According to the World Bank report (2015), in many African countries, only the upper 5 to 10 percent of the population can afford the cheapest form of formal housing. This implies that the remaining 90 to 95% of the population live in informal housing.

The Strain on Urban Services

The sudden surge in urban population due to migration places a heavy burden on essential services such as healthcare, education, and water supply. Cities on the

African continent are greatly affected by shortage of social services such as access to clean water. Similarly, due to the lack of toilets, the use of dirty water, and poor hygiene lead to increase in diseases. As a result, inadequate provision of these services negatively impacts the quality of life for both migrants and long-term urban residents.

Cultural and Social Changes

Migration can lead to cultural shifts and social changes in both rural and urban communities. Migrants often bring their cultural practices and traditions to the cities, contributing to the diversity of urban life. However, this can also lead to tensions and challenges as different cultural groups interact and integrate.

Rural Decline

As people leave rural areas, the communities they leave behind may experience a decline in population and economic activity. This can have negative consequences for rural economies, agriculture, and overall community dynamics.

Skills Drain

The migration of skilled and educated individuals from rural areas to cities can result in a "brain drain" effect. The loss of skilled workers can hamper rural development and limit the potential for growth in rural regions.

Remittances and Economic Impact

On the positive side, remittances sent back to rural areas by migrant workers can provide a significant source of income and contribute to local economic development. However, overdependence on remittances can also create vulnerabilities if the urban economy experiences fluctuations.

Environmental Impact

Urbanization and the concentration of people in cities can lead to increased pollution and resource consumption, impacting the environment. Additionally, the expansion of urban areas can encroach upon nearby agricultural lands and natural habitats.

Family and Social Dynamics

Migration can have complex effects on family and social dynamics. While it may improve the economic situation for migrants and their families, it can also lead to social disconnection and challenges in maintaining traditional family structures and support systems.

Urban Poverty and Inequality

Rural-urban migration can contribute to urban poverty and income inequality. While some migrants may find success in urban areas, others may struggle to secure stable employment and adequate living conditions, leading to pockets of poverty and disparity within cities.

Generally, as discussed above, rural-urban migration is a complex issue with far-reaching implications for both rural and urban areas. While migration can provide individuals with new opportunities and improved living standards, it also poses significant challenges to urban centers as presented above. Hence, addressing rural-urban migration requires a multi-faceted approach, involving investments in rural development and effective urban planning.

Consequential Effects of Urbanism in Africa

The on-going urbanism on the African content as noted from its features that we have discussed in the previous section has several effects. Most of these effects have a negative orientation as opposed to the positives. In this section, we will therefore look at some of the negative effects that are coming along with the ongoing urbanism on the continent.

Overcrowding:

This refers to the presence of more people or things in a particular space than expected. Ideally, it can also be seen as a situation whereby more people are located within a given space than is considered tolerable from a safety and health perspectives.

When urban areas become overcrowded, people start building slums, the roads become very busy with high traffic, public markets and malls. Poor air and water quality, insufficient water availability, waste disposal problems and high energy consumptions are exacerbated by the increasing population density and demands of urban environments.

In such situations, strong city planning is essential for responding to such challenges. (Odendaal, 2021). The picture below shows the state of overcrowding in Ghana.

Figure 8: State of Overcrowding in Kumasi City in Ghana

Development of slums:

A slum is a highly populated urban residential area consisting of densely packed housing units of weak build quality and often associated with poverty. The infrastructure in slums is often deteriorated or incomplete, and they are primarily inhabited by impoverished people. Although slums are usually located in urban areas.

In some countries they can be located in urban areas where housing quality is low and living conditions are poor. Slums differ in size and other characteristics most lack reliable sanitation services, supply of clean water reliable electricity, low enforcement and other basic services (Cinnamon, 2022). The picture below demonstrates a sample of a slum in South Africa.

Figure 9: Sample of a Slum in South Africa

Unemployment:

Occurs when someone is willing and able to work but does not have a paid job. Unemployment rate is the percentage of people in the labour force who are unemployed. As many people migrate from the rural areas to urban areas, cities become highly populated with no job opportunities to accommodate everyone in the city. As a results, many people find themselves unemployed on account to two factors. First, when the number of city residents is higher than the available job opportunities. Secondly, when the population from the rural areas lacks the necessary skills to be absorbed in the job market. The impact of high levels of unemployment is devastating, resulting in poverty, social exclusion, inequality, crimes and social instability.

Poor health and spread of diseases:

Urbanism as a product of urbanization has long been associated with human development and progress, but recent studies have shown that urban settings can also lead to significant inequalities and health problems. Cities are known to play multifaceted functions in all societies. They are the heart of technological development and economic growth of many nations, while at the same time serving as a breeding ground for poverty, inequality, environmental hazards, and communicable diseases (McMicheal, 2000).

When large numbers of people congregate in cities, many problems result, particularly for the poor. For example, many rural migrants who settle in an urban slum area bring their families and their domesticated animals both pets and livestock with them. This influx of humans and animals leads to vulnerability of all migrants to circulating communicable diseases and the potential to establish an urban transmission cycle.

Further, most urban poor live in slums that are unregulated, have congested conditions, are overcrowded, are positioned near open sewers, and restricted to geographically dangerous areas such as hillsides, riverbanks, and water basins subject to landslides, flooding, or industrial hazards. All of these factors lead to the spread of communicable and non-communicable diseases, pollution, poor nutrition, road traffic, and so on (Getaz et al 2011). The problems faced by the poor spill over to other city dwellers.

Some of the major health problems experienced in urban cities include poor nutrition, pollution-related health conditions and communicable diseases, poor sanitation and housing conditions, and related health conditions. All these have direct impacts on individual quality of life, while straining public health systems and resources in urban areas/cities (Kennedy, 2009).

High cost of living

As urbanism continues to increase in most of the cities on the African continent, the cost of living is equally increasing. In general, African countries have unique challenges when it comes to the cost of living. But at the end of the day, it all boils down to factors like economic policies, inflation, and the cost of essential goods and services.

These are the main culprits that determine how much it costs to live in an African country. If the economy is doing well, demand for goods and services will go up, which usually means a higher cost of living. And over time, inflation will drive up the prices of everything, making it even more expensive to live there. Unfortunately, the high cost of living can hit hard for those with fixed or low incomes, making it tough to make ends meet, let alone afford basic necessities or save for the future. This can also make it hard to access crucial services like healthcare and education as revealed by (Orufa, 2023).

On the flip side, a low cost of living is crucial for any country's economic stability. It affects the standard of living, purchasing power, and investment opportunities. Keeping costs low can help fuel economic growth, with increased demand leading to more production, employment opportunities, and a thriving economy. It can also make it easier for businesses to operate in the country by reducing their operational costs.

On a similar note, Galal (2023), revealed that Dakar, in Senegal is the most expensive city in Africa with an index score of 47.2. Other cities considered to be expensive include Addis Ababa (Ethiopia), Abidjan (Ivory Coast), Harare (Zimbabwe) and Johannesburg (South Africa) among others as shown in the figure below.

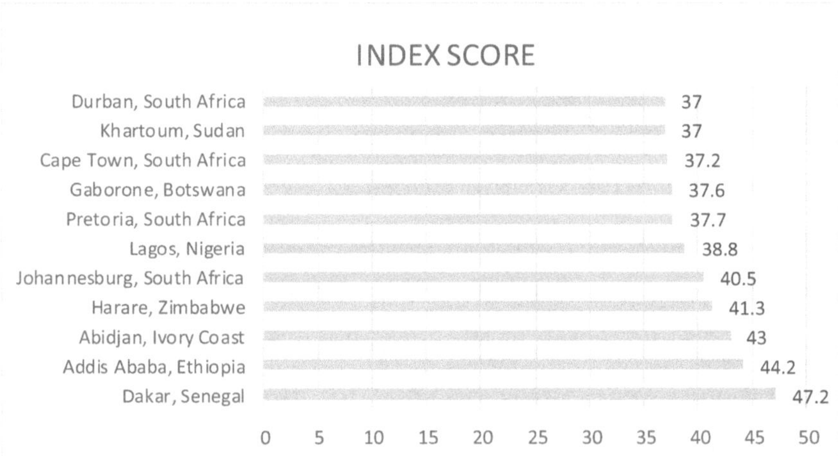

Figure 10: Index Score for High Cost of Living in Selected African Cities

Urban crime

As urbanism takes course in most African cities, urban crime as a resultant effect of this process is increasing as revealed by Dodman et al (2017). By using Numbeo composite crime indices, some of the most 10 dangerous cities in order of ranking as per Numbeo crime indices in Africa include the following;

1. Rustenburg, South Africa

Rustenburg City is in the North West Province of South Africa. It has very high crime levels, and is rated at 85.71 out of a possible crime score of 100, according to a 2015 Numbeo report. Of all of the cities in the North West Province in 2015, Rustenburg had the highest total cases of criminal activities with 11,117 cases in 2015, according to Crime Statistics South Africa. Household burglary, kidnapping, hijacking, and political violence are but some of the crimes that frequently occur in this city.

2. Pietermaritzburg, South Africa

Pietermaritzburg is both the province's second largest city (after Durban) and the capital of Kwa-Zulu-Natal Province in the nation of South Africa. It has very high crime levels, with a crime rating of 87.5 out of a possible crime score of 100, according to Numbeo as of January 2016. Armed robbery, sexual assault, arson, drug dealing, house break-ins, carjackings, and car theft are some crimes that are rife in Pietermaritzburg. In the 3 years leading up to 2015, crimes levels in the city rose significantly. According to Crime Statistics South Africa, in 2015 there were 15,720 criminal incidences in the city. This was an increase on the previous years, as in 2014 there were 14,794 reported incidents of criminal activities, and in 2013 there were 13,596.

3. Benghazi, Libya

Since Muammar Gaddafi's oust from power as the leader of Libya in 2011, the whole country, including the second largest city of Benghazi, has been engulfed in a civil war. The city is ever at risk of radical Islamic terrorist attacks. In September of 2012, the US Embassy in Libya was attacked by violent extremists, and 4 US government employees, the US Ambassador to Libya included, died in the chaos.

According to the OSAC, crime levels across the whole country are still rising. Incidences of carjacking, robberies, burglaries, and gun attacks are now more prevalent, largely due to the looting of government artillery that ensued in a mad scramble for economic and military gains after Gaddafi's removal. Sexual harassment of women is also rising, as well as petty and hardcore street crimes alike.

4. Johannesburg, South Africa

Johannesburg is South Africa's largest city, with an estimated population of over 4 million people, and it's also the capital of Gauteng province. In recent years it has achieved notoriety for the high numbers of crime and rape cases seen there as well. Numbeo ranks crime levels in Johannesburg as being very high, achieving a 91.61 out of a possible crime score of 100 as of March 2016. Due to the many incidences of rape reported there, it has also been dubbed as the "World's Rape Capital"

5. Durban, South Africa

Durban is the largest city in Kwa-Zulu-Natal Province in South Africa. It's a coastal city that is popular with local and international tourists who annually number within the hundreds of thousands. Economically, the city hosts the biggest container and commodity port in Sub-Saharan Africa, according to the OSAC. But crime and criminal-related deaths have been on the rise.

According to a 2014 report by the Mexican Citizens' Council for Public Security and Criminal Justice, Durban City was 38th among the 50 most violent cities in the world. Homicides per 100,000 people stood at 34.5 in 2014, up from in 2013 when they were at 32 per 100,000. As of March 2016, Numbeo rated the crime in Durban at 87.89 out of a possible crime score of 100, indicating it as being very high. Beyond Durban's city limits, Kwa-Zulu-Natal province in 2015 was reported as the most dangerous province in South Africa.

6. Lagos, Nigeria

With an estimated population of 21 million, Lagos is Nigeria's largest city and one of the largest cities in the world. Crime levels are also very high, with Numbeo giving the city a rating of 80.88 out of a possible crime score of 100. The OSAC gives Lagos a crime rating of "critical". Locals and expatriates alike living in Lagos are prone to muggings, armed robberies, home or business burglaries, carjackings, assaults, rapes, kidnappings, and extortion. Armed robbers in the city are known to even climb up perimeter fences and subdue guards, and others have invaded waterfront compounds by way of boats. Criminals also target vehicle occupants in traffic, and break into stationary cars.

7. Cape Town, South Africa

Crime levels in Cape Town, a port city in South Africa and the capital of Western Cape Province, are very high, even though it's the country's legislative capital. In recent years, it has consistently been ranked among the most violent cities in the world. Numbeo gives the city a crime rating of 82.45 out of a possible crime score of 10 as of

March 2016, which is an increase from the previous 3 years.

From April 2011 to March 2012, Cape Town recorded more murders than both Johannesburg and Pretoria combined, according to Africa Check. Criminal activities like drug dealing, mugging, vandalism, theft, assault, armed robbery, and bribery are especially common there, according to Numbeo. Taking nighttime walks is also risky in and of itself. Much of the violence in Cape Town is fueled by drug-related gang activity and the economic inequality still prevalent in non-white territories.

8. Port Elizabeth, South Africa

Port Elizabeth is a coastal city in Eastern Cape Province, South Africa. Tourists are drawn to this vibrant port city, which is dubbed the "Windy City" for its windy coastal clime. Port Elizabeth has a rich cultural heritage, and is among the cities early explorers like Vasco Da Gama passed through.

However, crime is very high there, and Numbeo, as of February 2016, rates the crime levels there at 80.56 out of a possible crime score of 100. Muggings and street crimes are common in Port Elizabeth much like in the rest of South Africa. Xenophobic attacks against foreigners also tend flare up in Port Elizabeth when there is unrest in South Africa. In 2014, Port Elizabeth was ranked number 35 by the Mexican Citizens' Council for Public Security and Criminal Justice among the world's 50 most dangerous cities, and, in 2013, it was number 41. Murders per 100,000 people in 2014 were 34.8, slightly down from 36 in 2013.

9. Luanda, Angola

Luanda, Angola's capital city, has a high crime rate, and one which the OSAC reports as being "critical". Out of a possible crime score of 100, Numbeo rates it at 76.39 as of February 2016. According to the UK Government service Gov.UK, common crimes in the city are inclusive of carjackings, assaults, homicides, muggings for valuables such as mobile phones, armed robberies at either night or day (especially in areas popular with foreigners), and rape incidences both in nightlife areas and even private homes.

Lone night travel or walks within the city are not recommended, as crimes mostly happen at night in Luanda, according to the OSAC. There also are crime hot-spots in Luanda that the OSAC particularly warns against venturing into, like the Rua Nehru, the Rua Houari Boumedienne, and the traffic circle in the last part of Rua Gamal Abdel Nasser.

10. Nairobi, Kenya

As Kenya's capital city, Nairobi is the hub of the country's economy, thereby making it attractive to both foreign and investments, and vulnerable to criminal activity. The US Overseas Security Advisory Council (OSAC) rates Nairobi's crime levels as being "critical". Also, Numbeo,com, as of March 2016, ranked the crime levels at of city as high, and ended up giving it a rating of 78.49 out of 100.

According to OSAC, serious crimes common in Nairobi, are armed carjackings, home and business burglaries, kidnappings, pick-pocketing, muggings, and snatch- and grab-thefts. Other criminal threats include grenade attacks and terrorism, which in recent years Kenya has bore the brunt of. However, those victims of robbery who cooperate without resisting are likely to be unharmed in the city, according to OSAC. Cyber-crime and credit card skimming are other forms of criminal activities on the rise in Nairobi that its foreign visitors need to be aware of.

Poverty

Poverty is a state or condition in which a person or community lacks the financial resources and essentials for a minimum standard of living. In such instances, poverty-stricken people and families sometime go without proper housing, clean water, healthy food, and medical attention. Arguably, as urbanism takes place in urban areas, poverty with all its related consequences also increases in most cities in Africa.

Notably, this challenge is much felt in secondary cities on the continent. As stressed by Roberts (2014), secondary cities may include cities with a population of 100 thousand to 5 million, although they may be larger or smaller depending on a country's total population. In addition, Roberts (2014) argues that from 2010 to 2025, secondary cities will host about 460 million of global urban population, compared to 270 million in megacities. This situation will undermine efforts towards sustainable urban development in these cities.

Traffic congestion

Traffic congestion is a major concern for many cities in Africa. Many African cities are growing at an exponential rate and experience huge traffic that they cannot accommodate.

Precisely, traffic congestion is a common sight in bustling city areas, where vehicles, pedestrians, and public transport coexist in a complex web of movement. Such traffic is attributed to several factors that may include; high population density, inadequate infrastructure, limited public transportation, and urban sprawl, amongst other factors.

Additionally, it is important to note that traffic congestion is not merely an inconvenience; it has far-reaching consequences, such as economic costs, environmental implications, physical and mental health deterioration, and others. It is therefore a multifaceted issue that requires a comprehensive approach. Hence, by utilizing tools like the Traffic Index, city planners can gain valuable data to develop effective strategies for reducing congestion, improving public transport, and promoting sustainable urban mobility.

Hence, based on Numbeo's Time Index which is a platform that delivers real-time information on the cost of living, quality of life, and other socioeconomic indicators in cities and nations throughout the world, Africa's five most affected cities by traffic congestion are as presented in Table 2 below.

Table 2: The Five African Cities with the worst Traffic Index

Rank	*City*	*Country*	*Traffic Index*	*Time Index*
1	*Lagos*	*Nigeria*	*342.5*	*67.4*
2	*Nairobi*	*Kenya*	*253.6*	*53.4*
3	*Cairo*	*Egypt*	*244.5*	*50.2*
4	*Pretoria*	*South Africa*	*230.5*	*45.5*
5	*Cape Town*	*South Africa*	*208.9*	*42.5*

Urbanization as a Major Driver of Urbanism in Africa

Urbanism is an outcome of the urbanization process. This implies that when we begin to question ourselves on what drives urbanism in Africa, our very stating point is the urbanization process. Other debatable drivers may include rate of economic growth, political influence, instituted urban policies and regulations among others. In the context of this books, we will dwell more of urbanization as the founding factor for urbanism.

Ideally, no country in the world has ever reached middle income status without a considerable population shifting into cities. Similarly, we probably know when we say 'I am living in a city or I am moving to a city'. The anticipation of a better life builds up in our minds to the extent that even some people tend to believe that if you have not been to a city, you have not lived life to your fullest. Having these thoughts in our minds is a huge trigger of many people migrating to cities from rural areas. Such movements culminate into what we could call urban movements and thus urbanization.

Generally, the share of total population on the African continent shifting to cities is gradually increasing as depicted in the figure below.

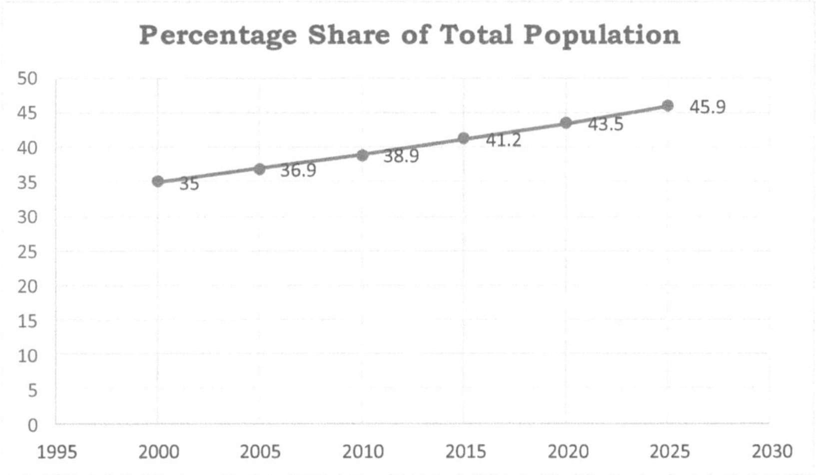

Figure 11: Share of Total Population Shifting to Cities on the African Continent

Source: Statista, 2023

In fact, the urbanization process has come to be labelled as one of the most profound transformations that the African continent will undergo in the 21st century (Afdb, 2022). As stressed by the African Development Bank, since 1990, the number of cities in Africa has doubled in number from 3300 to 7600 and statistics as revealed by the United Nations, Department of Economic and Social Affairs, Population Division (2019) show that Africa is undergoing an unprecedented urbanization phenomenon. Since 2010, the urban population of Africa has grown by 21 million every year. In 2015, for example, Kenya had more urban dwellers than the whole of Africa in 1950.

However, it should be noted that since the 1990s, strong demographic growth has been the primary driver of this urbanization. Africa's population has more than doubled since 1990 to reach 1.2 billion people in 2015, and is set to double again by 2050. These urbanization dynamics on the African continent are influenced by many other factors whose impacts vary depending on the national context: climate and geography, people's income, public institutions and policies, economic cycles, conflicts, etc. Below is the list of the top and lowest ten countries with their urbanization rates as revealed by Galal (2023).

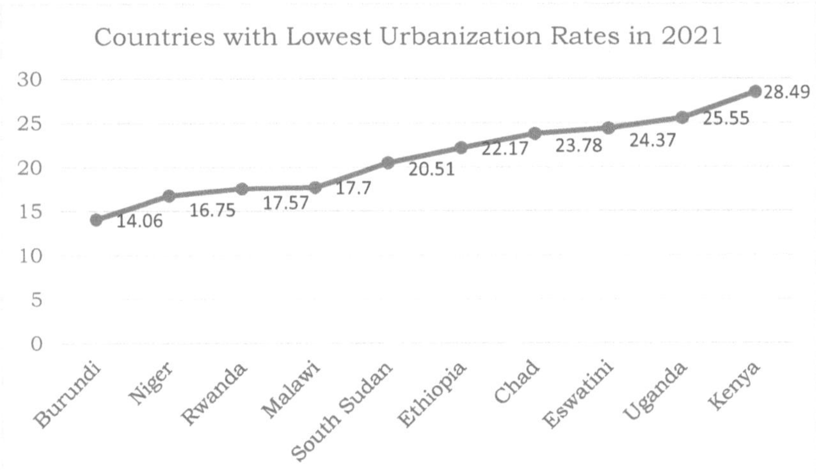

Figure 12: Lowest Urbanization Rates in 2021

Source: Galal.S (2023)

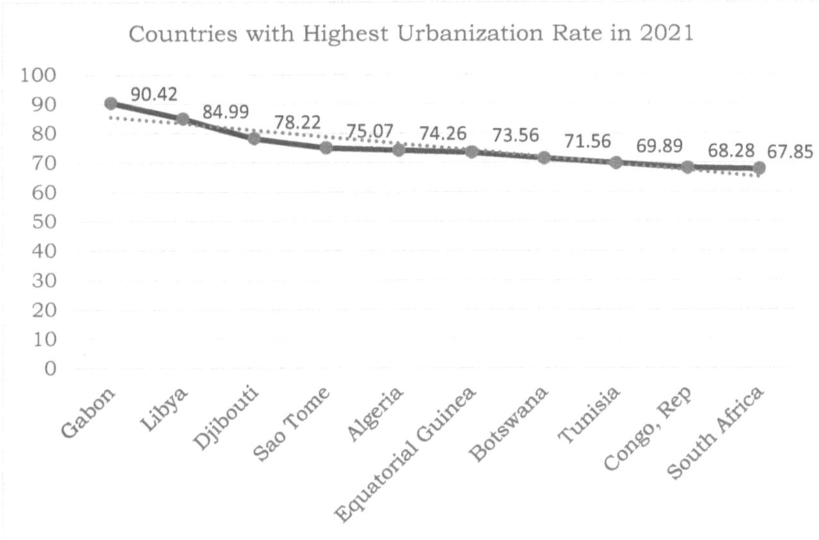

Figure 13: Highest Urbanization Rates in 2021

Source: Galal.S (2023)

Generally, from the above two graphs, Burundi is seen as the lowest urbanizing country and Gabon the highest in the year 2021. Nevertheless, the ongoing urbanization raises concerns among various stakeholders. For instance, according to Collier (2016) in a policy brief focusing on African Urbanization, Collier argues that the ongoing urbanization is dysfunctional, the key indication being that cities are not generating enough productive jobs and those moving into cities lack the necessary skills to match with the employment needs of cities. Hence, if urban policies remain unchanged, future urbanization is likely to result in similar outcomes.

PART FOUR

SUSTAINABLE URBANISM AS A NEW URBAN AGENDA FOR AFRICA'S RAPIDLY URBANIZING CITIES

Overview

Part four of this book focuses on guiding the reader into understanding the concept of sustainable urbanism as a New Urban Agenda for Africa's rapidly urbanizing cities. In this part of the book, the writer acknowledges that urbanism can occur in any urbanized area but the overall goal should not be to simply experience urbanism, rather, to experience sustainable urbanism.

The Concept of Sustainable Urbanism and Its Evolution

Looking back in time, sustainability was never a central theme in urbanism. The famous image of the architect eating the landscape, by Malcolm Wells, illustrates the way urbanism was seen for a long time: as a destroyer of natural values. Girardet (1996) was one of the first to acknowledge the importance of an integrated approach to developing cities in a sustainable way. He connected themes such as looking at cities as ecosystems, the footprint of cities, urban heat islands, and others with urban design schools, such as the garden city movement

(Howard 1902), the modern city of Le Corbusier (Cohen and Benton 2008), Frank Lloyd Wright's Broad acre city (Wise 2013) and megacities in general (Nicholls 1995; Donatiello 2015). His pledge for convivial cities the role of citizens and the local scale and the attention for health, greening the city, renewable energy and recycling, and sustainable forms of mobility, are still relevant. Since then many perspectives, visions and conceptual practices have been developed, each of them taking sustainability into account in planning and designing cities in one way or another.

Hence, since the 'Brundtland report' (World Commission on Environment and Development 1987) sustainable development has become the main goal for many sectors including the urban policy-making process. This has also impacted the practices and theories of urbanism. Over the years, this has led to many frameworks, concepts and planning principles being established. One often noticed outcome of this goal is the Freiburg Charter (Daseking et al. 2012), in which the principles are defined in three groups: spatial development, content and process. These principles provide guidelines on how a sustainable city can be developed.

Sustainable urbanism is defined as "walkable and transit-served urbanism integrated with high performance buildings and high-performance infrastructure" (Farr 2007). A broader definition is: "the application of sustainability and resilient principles to the design, planning, and administration/operation of cities" (Sharifi 2016).

Compactness (density) and biophilia (human access to nature) are considered as the core values of sustainable urbanism (Adhya et al. 2011). Others question compactness as being really sustainable (Daneshpour and Shakibamanesh 2011; Neuman 2005). In practice, the ten leaders in urban sustainability each have their own thematic focus area: transport in Bogota, energy in Melbourne, air quality in Mexico City and so on (Schwartz 2013). Hence the main focus of sustainable urbanism lies on the design of urban environments and aims to implement sustainable solutions through these designs. The leading cities however show strong preference for certain themes.

In order to become sustainable cities, need to adopt a new approach to planning the city. In the World Cities Report (UN Habitat 2016) this is described as the city that plans in contrast to the planned city. The 'city that plans' uses integrated and multi-sectored planning approaches, takes into account local circumstances and involves diverse populations including women, plans are prepared at different geographical scales and across political boundaries, and improves the education for planners, especially in developing countries.

A city that plans not only projects the future from the past trends, it also brings the public, private and third sectors together with communities to build a collectively preferred future. These aspects of planning don't seem ground-breaking, but in many cities they are not applied. Hence, in a city that plans, a number of principles are defined to support implementation.

Such principles include; promote sustainable development, achieve integrated planning, integrate plans with budgets, plan with partners and stakeholders, meet the subsidiarity principle, promote market responsiveness, ensure access to land, develop appropriate planning tools, be pro-poor and inclusive, and recognize cultural diversity (Farmer et al. 2006). By adopting these principles, such a city will be able to harvest the benefits of the New Urban Agenda.

In principle, the New Urban Agenda aims at harnessing the potential of cities and human settlements to help eradicate poverty in all its forms and dimensions, reduce inequalities, promote inclusive growth, and achieve sustainable development (Habitat III 2016). Hence it provides a new global framework to develop sustainable cities. One of the principles defined is to ‘transform the way we plan, develop, govern and manage cities and human settlements, recognizing sustainable urban development as an essential instrument to achieve prosperity for all and sustainable development.

Necessity for Sustainable Urbanism in Africa

As discussed in part three of this book, most experience urbanism that is largely dominated by small to medium cities with weak economies. As the author of this book, I find that there is room for African cities to shift from the urbanism paradigm to the sustainable urbanism on account of three major reasons.

First and foremost, shifting from urbanism to sustainable urbanism will enable African cities to become resilient to negative environmental related effects such as disasters, climate shocks and disease out breaks. It should be noted that sustainable urbanism advocates for building cities that respond to the principles of sustainability, livability and accessibility among others. By so doing, cities are meant to be built on the foundations of these principles which will guarantee their ability to resist disease out breaks, climate change shocks and any disaster.

Secondly, sustainable urbanism is of necessity to African cities because it will assist in reducing social inequalities among city residents. This is based on the fact that sustainable urbanisms advocate for inclusivity and diversity. As a result, sustainable urbanism tends to celebrate diversity in all its forms, including cultural, economic, and social diversity so as to promote inclusivity and a sense of community. This will also enhance prosperity for all and thus give an opportunity to all city residents' equal chances of living a social justice life.

Thirdly, sustainable urbanism is of necessity to African cities because it will enable African cities to become engines of economic growth. As discussed earlier, the current weak economies manifested in most African cities don't give them the opportunity to serve as engines of economic growth. This is because most economic activities in these cities are small and heavily struggling to survive. However, with the adoption of sustainable urbanism which advocate for sustained growth, cities will be in position to come up with strategies that solidify their positions as engines of economic growth.

Principles of Sustainable Urbanism

As a process that manifests a way of urban life, sustainable urbanism is based on a set of principles that govern its manifestation in urban areas. These principles align with the political, social, economic, environmental, cultural and political aspects of urban areas. Hence, scholars such as Gibbs, (2012) categories these principles to include the following;

Sustainability: Sustainable urbanism should promote sustainable development by minimizing environmental impact and maximizing efficiency in the use of resources. Sustainable urban development is defined as development that meets the needs of a community without compromising the ability of future generations to meet their own needs.

It prioritizes economic, environmental and social sustainability. Sustainable development aims to ensure that cities are live able and supportive of economic, social and ecological resilience in the face of a changing climate. Sustainable development focuses on creating cities that are not only economically vibrant but also include a quality of life that is balanced between the social and environmental needs of the communities. Sustainable development combines strategies to reduce pollution and preserve natural resources with improvements in transportation and housing, public safety and health, and social amenities.

It includes integrated approaches to urban planning that are responsive to the needs of local stakeholders, sustainable urban development is so essential is the economic benefits it can bring. By investing in sustainable initiatives and practices, cities can make themselves more attractive to businesses, drawing in new enterprises and creating jobs. Utilizing energy efficiency strategies can reduce operating costs and make businesses more profitable, allowing them to offer better wages, which in turn boosts the local economy.

Additionally, reducing the amount of waste produced by cities can bring down the cost of municipal services, freeing up funds for other essential services, such as healthcare, education, and housing. Finally, sustainable urban development is essential for ensuring that all citizens can enjoy economic work and better quality of life.

Livability: Sustainable urbanism should enhance the quality of life for residents by providing safe, healthy, and appealing environments that support social interaction and community engagement. Urban livability is the ability of urban spaces to fulfill the expectations of its inhabitants for wellbeing and quality of life. Measurable spatial patterns underlie the emergence of livable cities. Still, few researchers have considered if and how these patterns affect socioeconomic conditions across spatial scales.

Urban livability can simply be rephrased as quality of life in cities, but its concept is complex because of the heterogeneous nature of cities and the encompassing meaning of quality of life. Urban geographers focus on human behavior–environment relationships, and define urban livability as the extent to which a city can satisfy the physical and psychological needs and demands of its residents.

Geographers assess residents' satisfaction by using both objective and subjective indicators and information, and explain live ability by delving into contextual factors. The common objectives of urban planners and designers are to make cities vital, to enable them to impart a sense of place and time, to ensure that they fit well with human activities, to make them accessible, controllable, and able to attract continuing investment. With the advent of neoliberalism and globalization, global livable ranking systems emerged in the 1990s. These relay the views and ratings of city-dwellers, visitors, and other stakeholders, supplemented by secondary information.

The livability of high density cities or districts characterized by high rises has become a major contemporary concern following the pursuit of sustainable urban development. It is expected that two development strands, the global aging trend and the advancement in smart technology, will impact on the future directions of livability research.

Accessibility: Sustainable urbanism should provide easy access to jobs, services, and amenities for all residents, including those with disabilities and those who cannot afford private transportation. Accessibility means that all people can perceive, understand, navigate, and interact with electronic information and be active, contributing members of the digital world.

Access to people, goods, services, and information is the basis of economic development in cities. The better and more efficient this access, the greater the economic benefits through economies of scale, agglomeration effects, and networking advantages. Cities with higher levels of agglomeration tend to have higher GDP per capita and higher levels of productivity. The way in which cities facilitate accessibility through their urban forms and transport systems also impacts directly on other measures of human development and well-being.

Urban travel currently constitutes more than 60% of all kilometres travelled globally (Korniychuk et al. 2014) and, as a result, urban transport is currently the largest single source of global transport-related carbon emissions and the largest local source of urban air pollution.

Inclusivity and Diversity: Sustainable urbanism should celebrate diversity in all its forms, including cultural, economic, and social diversity, to promote inclusivity and a sense of community. The key to improving the quality of city life is creating "urban diversity." This means promoting diversity in the types and sizes of businesses and business spaces, in cultural activities and institutions as well as economic diversity.

Like genetic diversity in the natural world, diversity in an urban context builds resilience throughout the city's systems. What are the Benefits of Diversity? Cultural enrichment which results from learning about other cultures helps us to understand other viewpoints and ideas and in doing so enriches our lives.
Greater creativity that involves having a variety of cultures and backgrounds allow for different perspectives and new ideas. This diversity of thoughts and opinions promotes creativity and innovation. Hence, cities with a diverse population tend to be more innovative than homogeneous ones. Better lifestyle with people from different backgrounds come a myriad of different traditions, cuisine, and art.

Finally, greater civic engagement which involves the willingness of citizens to get involved in their communities to strengthen them and make them a better place for everyone to live and work.

Compactness:

Sustainable urbanism as revealed by Bibri et al., (2020) should promote compact, walkable, and mixed-use development to reduce sprawl and the need for private transportation, thereby reducing emissions and promoting healthier lifestyles.

Hence, if compact cities could be seen to be safer and more resilient, there would be direct policy implications for how cities should grow and/or rebuild in the face of increasing hazards associated with global climate change.

In addition, there is no precise level of density specified for smart growth, but those areas touted as smart growth or new urban developments will often have densities as high as 10,000 people per square mile, with some as high as 20,000 people/sq mi or even much higher, versus 3,000 – 4,000 people/ sq. mile in typical suburban developments.

Many of the benefits that accrue from smart growth devolve from this increased density. Interestingly, a common planning recommendation in coastal hazardous areas is to reduce, rather than, increase density. From the environmental perspective, the most obvious benefit of compact growth is less open space is consumed by development. Similarly, proximity and compactness, if well designed, confer a measure of urban vitality not achievable with conventional car-dependent, sprawling development.

Resilience:

Sustainable urbanism as stressed by Chenwei et al (2023) should promote resilience in the face of natural disasters, social disruptions, and economic challenges, by creating flexible and adaptable urban environments that can respond to changing circumstances. Urbanization, combined with the increasing frequency and intensity of natural hazards associated with the onset of climate change, is contributing to rising risk exposure in the Pacific. The region's cities and towns serve as hubs for administration, essential social service provision, and drivers of economic growth. However, because of this concentration of people, infrastructure, and economic activity, urban centers are highly exposed to natural hazard and climate change risks.

Many of the region's urban centers are located in hazard-prone areas such as coasts, flood plains, or low-lying atolls. Urban resilience is the ability to withstand and recover from unexpected shocks associated with natural hazards and climate change. Building resilience to these risks can be dealt with most effectively through strengthened urban planning and management.

Considering and responding to these risks is an essential part of proactive urban management strategies and infrastructure development plans. Possible approaches for building urban resilience include the following: preserve natural ecosystem functions such as natural drainage channels, green space, and natural shoreline buffers; and implement effective early warning systems, emergency disaster response, and post-disaster recovery.

Conservation and preservation: Sustainable urbanism should emphasize the responsible use and protection of natural resources. Conservation involves the sustainable management of resources to ensure their long-term availability, while preservation aims to protect and maintain natural ecosystems and biodiversity in their original state. For example, in Tanzania the government address policy of using alternative source of energy such as the uses of gases instead of charcoal so as to preserve forestry.

Interdependence: Sustainable urbanism should recognize the interconnectedness of various elements in the environment, including living organisms, ecosystems, and the physical environment. It highlights the importance of considering these relationships and the potential ripple effects of environmental actions.

Equity and justice: Sustainable urbanism should emphasize the fair distribution of environmental benefits and burdens across different populations and communities. Environmental decisions and actions should consider the needs and rights of marginalized and vulnerable groups, and strive for environmental justice

Social Cohesion: Cultural perspectives on urbanism often prioritize social cohesion and community engagement. Hence, Sustainable urbanism should emphasize the importance of creating inclusive and equitable urban spaces that foster social connections and a sense of belonging among residents. This can involve designing public spaces, neighborhoods, and infrastructure that promote interaction and community participation.

Cultural Identity: Sustainable urbanism that is guided by cultural perspectives recognizes and celebrates the unique cultural identity of a place and its inhabitants. It seeks to preserve and promote local traditions, heritage, and cultural practices through urban design, architecture, public art, and cultural programming. This approach helps maintain a sense of continuity and pride in the community's history and identity. For example, the

Maasai in Arusha, Tanzania have preserved their cultural identity through their wearing styles.

Strategies for Enhancing Sustainable Urbanism

Strategy One: Fixing the future

The strategy to fix the future aims to increase the environmental resilience of the urban environment. The future is repaired, even before it is lost. Through the application of urban design principles, the loss of biodiversity, resources and environmental quality is limited and compensated for. In and around the city ecological water and green systems are implemented and renewable energy sources are cultivated or generated. The mobility is minimized and sustainable modes of transportation are implemented. In and around the urban settlements a range of food production is planned to provide the urban population with fresh, naturally produced crops.

The resources used and also the waste produced during processing and use in the city are minimized. An analysis of the diverse flows, which come into the city, are processed and re-used or put through the city and then leave the city again, gives insights in the amount and size of spaces required for these flows to be accommodated and treated, processed or distributed. This analysis and these insights, in combination with existing green and water spaces, forms the input and condition for designing a sustainable city.

There are many examples already of these so-called eco-cities or eco-villages designed and realized around the world, such as Morra Park in Drachten (Beatley 2000; Kristinsson 2012), Westerpark in Breda (Gemeente Breda 1993) all in the Netherlands and Freiburg in Germany (Freytag et al. 2014) among others.

Hence, the mainstream definition of sustainable urbanism is along these lines of thinking that include; repairing the harm done to the natural system, recycle and reuse resources, and clean waste. When the urban system is designed to foster sustainability in this way, new uncertain developments, such as the impacts of climate change, are hard to incorporate.

When abundance of space is created for regeneration functions, green and water, urban agriculture and other 'environmental' functions, the chance is bigger this will satisfy the needs of unexpected changes. However, because these changes are inherently uncertain, the available space might be in the wrong place. In fact, this may lead to an unsustainable situation in the future and adaptive urban management (Birkmann et al. 2010) is required.

Enforced uncertainties are in principle easy to plan for. These new uses require space in urban areas, but if these transitions link up with sustainability principles, they are able to co-exist in the same areas. An example of this is the situation where windmills are combined with natural wetlands.

Locating new urban developments in vulnerable areas increases the exposure to uncertainties, hence this is not recommended as these spaces are required for creating the space to implement sustainability principles. For instance, building in the coastal zone minimizes biodiversity, leaves less space for regeneration of the natural system (cleaning the water, etc.) and exposes more housing to the risk of flooding. This could be arranged through strict regional planning, but in practice this is often either not possible or planning directives are too weak to prevent urban occupation of these areas.

Strategy Two: Indulging the future

The strategy of indulging the future aims to increase the spatial resilience in the city. In accommodating change urban form should be flexible and adaptable. Rather than creating fixed patterns, urban forms are allowed to emerge, the shape depending on external factors and change.

For the urban environment to be open to change a systems approach is helpful. Complex, self-organizing, adaptive systems, as cities are seen (Andersson 2012; Ikeda 2004; Bettencourt 2013, 2015; Batty 2013; Roggema 2012), have the capability to adapt to change.

Self-organization of the city, or the society of the city can be 'planned' by setting rules about the relation between built elements, being buildings, linear elements or district elements. These urban codes regulate how these elements relate to each other, informing planning in a qualitative way (Portugali 2007).

Urban coding will generate an indirect order, e.g. a nomocratic approach), by creating a stable and simple set of abstract and general relational rules, which enable society itself to be highly flexible. Leaving options open for society. The problem of complexity and intrinsic unpredictability requires a radical overhaul of the tools used to regulate land-use particularly by exploring the idea of shifting the emphasis to framework instruments, such as urban codes (Moroni 2015).

Swarm planning (Roggema 2012) suggests such a framework for dynamic environments, in which different paces of change are accommodated by different layers of planning, and planning starts with the biggest uncertainty or threat in mind. The impact of an intervention dealing with this uncertainty, creates a potential tipping point, which in turn regulates the spatial configuration of the future as it subsequently will be filled with spatial additions, and built elements by society.

Moreover, when built elements of which the city is comprised become dismantable, the urban codes can be reinterpreted and the city may self-organize into new, best-fitting configurations. In order for this to happen, spaces, and also physical elements, have to be able to change over sometimes very short time periods. Open spaces and structures in the city, such as the green network, should be designed in such a way that they can expand and shrink in size or change land use whenever necessary (Roggema 2014). This is a new task for urbanism.

The strategy of indulging the future is very well equipped to deal with uncertain developments, as the designed system encapsulates the potential to anticipate and increase resilience when an unprecedented event occurs, for instance when the capacity, flexibility and agility of the urban system are increased, through creating spaces for temporary land use, or 'unused' open spaces.

By using the strategy to indulge the future, new, unused or temporary spaces will be created to accommodate sudden changes. For instance, if an area is created along the coast for a sudden flood, this area will not be occupied by urban extensions and water could enter this place without destroying inhabited areas. However, if urban areas expand organically and anarchically, it is hard to keep areas free of occupation hence exposure of urban precincts to risks might increase.

PART FIVE:

INSTITUTIONALIZING SUSTAINABLE URBANISM IN AFRICA'S RAPIDLY URBANIZING CITIES

Overview

By 2050 the world's population is projected to reach 10 billion people, of which 67% is expected to live in urban areas, which is an increase from a 54% share of urban dwellers in 2016 (UN, 2018). In addition, the ongoing urban transition is increasing pressure on the earth's resource base and thus hindering sustainable urbanism through the generation of social negative externalities and diseconomies of agglomeration (Romero et al, 2011).

This part of the book therefore attempts to dig into the process of institutionalizing sustainable urbanism in Africa's rapidly urbanizing cities. Specifically, the author attempts to suggest possible ways in which sustainable urbanism could be realized elaborated below;

Adoption of the principle of inclusivity in Planning

Sustainable urbanism needs inclusive planning where all stakeholders are part of the process. It should be noted that since the Copenhagen Declaration (1995) and the First UN Decade for Eradication of Poverty (1997) and the Millennium Declaration (2000), global and national calls for creating a just and inclusive world that addresses the concerns of the weak and vulnerable have grown louder and planning for the goal has become pronounced through national multi-year plans, PRSPs and MTEFs and other home grown instruments around the world.

It is also understood from the various assessments that although there has been increasing attention to hear the voices of people and various stakeholders in programming at the national and sub-national level, the most excluded group like the older people, disabled, indigenous people and other groups like the youths are still not fully represented in the process and therefore the inclusive planning process still remains elusive.

This is particularly important because these vulnerable social groups constitute a large segment of the total population as demographic transition results in both ever increasing older people and youths in most countries. To this effect, PRSPs and national development plans need to find a way to include such excluded groups if they are to be inclusive to achieve the goals of creating a 'society for all' that not only aims for the higher purpose of social integration and equalization of opportunities for societies to accept all but also advocates for sustainable urbanism.

In order to realize inclusive planning so as to achieve sustainable urbanism, the following ought to be taken into account;

- ***Identifying the Social Groups and Weaker Sections***

Social Inclusion as defined by UNESCO (2015) refers to the process that enables equal opportunities everyone regardless of background. It is not confined to economic opportunities only and transcends to all aspects of human life to ensure individuals can realize the full potentials of their life. This is particularly important for vulnerable groups and weaker sections in society. Therefore, it is important to have identification of the groups and data for youth, disabled, older people and indigenous people and groups in country context (like disadvantaged castes, ethnic minorities etc.)

- ***Analyzing the Current Legal and Regulatory Environment***

Although social exclusion is often repudiated and rights of all citizens are usually proclaimed in no uncertain terms in all constitutions inclusion of all in the political process and every one's participation in the economic sphere is by no means assured through these constitutional guarantees. Appropriate examination of the legal and regulatory framework is therefore required to ensure there are no structural impediments for the usually excluded groups. For planning purposes, it is important that both the Government and city Systems analyze the existing legal and policy instruments to ensure that planning for all can be carried out without legal or regulatory hindrances and appropriate policy support is present to support inclusive planning.

- ***Promoting Participatory Methods in Policy Development and Planning***

Examine the practice of participatory methods in national/regional policy making and planning exercises. If there are precedents of employing participatory methods then the chances of inclusive policy making and planning may not be a distant dream, although for various reasons cited earlier it is by no means an automatic process, especially as it concerns less visible vulnerable groups

- ***Align concerns of Social Groups to National Targets***

Align the concerns of the vulnerable social groups to the national goals and targets particularly in line with global commitments and goals, e.g., Copenhagen Declaration, MDGs, ratification of global conventions on disability, non-discrimination, and rights of older people, indigenous people and youths. This is particularly important because while aggregate progress on key indicators may rise this can still mask the intra-regional, intra-social group disparities which will negate overall national progress.

Adopt Effective financing mechanisms

As stressed by UN-Habitat (2023), cities are the engine of national economic development and global development. Hence, ensuring that effective financing mechanisms to ensure that cities experience sustainable urbanism is of paramount importance.

Unfortunately, most African cities have extremely weak capacities to generate own source revenue, while facing the fastest urban growth. For example, in Iwo in Nigeria, the city's own source revenue only accounts for 2.2% of the total revenue. Kenema city in Sierra Leone can only raise a total revenue of USD 0.31 per capita, while the city of Aberdeen in UK with a similar size of population to Kenema raises a total revenue of USD 5,612 per capita. The revenue per capital in Aberdeen city is 18,103 times than that of Kenema city. This is an incredible contrast of revenue- generation capacity between African cities and cities in developed countries.

This makes it difficult to invest in key infrastructure needs of cities to meet the on-going urbanism. For instance, according to UN-Habitat (2023), there is a huge gap in infrastructure investment. Ideally, developing countries need an extra USD 1.3 trillion of investment in urban public infrastructure each year. The global infrastructure investment gap is USD 1.1 trillion to USD 1.5 trillion in developing countries. The infrastructure investment gap is around USD 180 billion a year for the Asia Pacific region, USD 24 billion a year for Latin America, and more than USD 93 billion a year for Africa.

For fragile African states, more than 37 percent of GDP needs to be invested in infrastructure. The global need for infrastructure investment ranges from USD30 trillion to USD40 trillion in the next two decades (Habitat report, 2023).

On account of the above situation, it is evident that responding to the on-going urbanism in cities of most African countries will only be possible if these cities can devise effective financing mechanisms that will enable them respond to development needs such as infrastructure and other related social services. To attain this goal, cities should;

- Diversify their funding sources and innovative financing instruments are two cornerstones for financing and delivering sustainable urban development.
- City governments should be empowered to raise their own resources to increase their revenue.

- ❖ Promote New Partnership Models to Leverage Private Sector Finance.
- ❖ Make effective use of external sources of finance such municipal bonds, climate financing mechanism etc.

Set up a monitoring, Evaluation, reporting and review mechanism

Collecting and maintaining high quality and up-to-date local data and information systems is a crucial responsibility that is fundamental for identifying problems and prioritizing policies, programs, and sustainable urban development interventions. In fact, at this point, we intend to respond to a key question on whether we are doing the right thing, and is our plan working?"

Ideally, holistic monitoring and evaluation allows government officials and development managers to learn from past experiences, improve service delivery, plan and allocate resources, while also demonstrating the results of the initiatives as part of their own accountability.

Hence, the city's potential to measure, monitor and evaluate the implementation of the New Urban Agenda and the progress of the strategic urban development objectives that respond to the on-going urbanism depends on its capacities to collect, process and analyze statistical as well as geographic data for informed decision making.

This implies that in order for use to measure how we are doing in terms of driving the new urban agenda, the cross-cutting processes of citizen consultation and stakeholder engagement at each stage must be accommodated. According to initiatives such as the UN-Habitat City Prosperity Initiative, the most prosperous sustainable cities are the ones that perform equally well in all dimensions of sustainability and that have successfully integrated planning, governance, and finance (UN-Habitat and International City Leaders 2015).

Improved coordination mechanism

Strategic coordination mechanism is a serious challenge in many African cities. As a results, the prevalence of this challenge undermines any efforts towards sustainable urbanism in these cities. Richards Veryards defined 3 key symptoms affecting coordination in any system and these include; delays, duplication and confusion with consequences including missed opportunities and inefficiency.

Depending on the level of coordination, macro or micro, different coordination mechanisms can be applied. Henry Mintzberg identified five coordination mechanisms. Firstly, mutual adjustment mechanisms where formal and informal discussions should be made between stakeholders in the city development process, secondly, direct supervision, where a manager or coordinator monitors activities. Thirdly, the standardization of skills where staff should be hired with standardized skills, such as bachelor or master degrees. Fourthly, under the Standardization of process mechanisms all necessary procedures, rules and regulations, specifying the contents of activities should be defined clearly and finally the standardization of, output mechanism products and services.

Generally, in winding up this part of the book, as the author, I wish to stress that institutionalizing sustainable urbanism in Africa's rapidly urbanizing cities could take some time considering the rate of rapid urbanization being experienced by these cities. Nevertheless, taking this direction by any city will be the most feasible way of establishing livable and sustainable cities on the African continent.

PART SIX:
CONCLUDING REMARKS

The process of developing sustainable urban cities needs to consist of the two strategies mentioned earlier in part four of this book. Such process should accommodate both the need for regeneration of vital functions within the urban realm and, in parallel as well as the creation of enough flexibility in order to accommodate sudden changes of land use. Additionally, it should create its urban environments in a way that makes them more resilient when something unexpected occurs. This triple-layered search for the sustainable and resilient city leads to a redefinition of sustainable urbanism.

Similarly, developing sustainable urban cities will require the designing of a sustainable urban system, which creates physical and mental space to adjust the urban form at any moment in time, anticipates uncertain, unexpected and unprecedented change, and grows stronger and becomes more resilient when uncertainty impacts on it. This implies that in the city, open spaces need to be created, which are not determined for one single use, where adjustments over time are possible and flexibility in land use can be easily and instantly achieved. Urban areas should be able to not only change function, but also change shape, acting as a breathing organism. Sometimes the city inhales people, when population grows, and sometimes it exhausts them, in times of shrinkage. Similarly, the city sometimes inhales water, in periods of heavy rain or a sudden flood, before slowly exhausting the water again.

For many functions, the city operates as a sponge, taking up and getting rid of materials, flows and organisms. In the urban design this requires the creation of spaces that are adjustable. At least 30 % of the territory should be available for temporary use or able to change its use rapidly as noted by Roggema (2012) and Broess (2014).

In addition, steering towards the path of sustainability and resilience in urban environments will greatly depend on effective institutions, governance and strategic planning within cities themselves. This is because rapid urbanization confronts city governments with unprecedented governance, planning, and fiscal challenges across the spectrum of urban sustainability. Hence, how decision makers prepare for rapid urbanization is crucial, not only to the future of the cities, but also to global economic progress and sustainability.

To this regard, City leaders must urgently develop visions for the future of their cities in consideration of their unique path for economic growth, environmental protection, climate impact and resilience, and inclusiveness. Such vision must be built upon integrated planning that utilizes a multidimensional framework.

Bibliography

Adhya A, Plowright P, Stevens J (2011): Defining sustainable urbanism: towards a responsive urban design. In: Proceedings of the conference on sustainability and the built environment. King Saud University, Saudi Arabia.

AFDB (2012): Urbanization in Africa. African Development Bank Group

African Development Bank (2022): Africa's Urbanization dynamics 2022: The Economic Power of Africa's Cities.

African Economic Outlook Report (2018): Africa's Infrastructure: Great Potential but Little Impact on Inclusive Growth.

Alirol E, Getaz L, Stoll B, Chappuis F, Loutan L (2011): Urbanization and infectious diseases in a globalized world. The Lancet infectious diseases.

Beatley T (2000): Green urbanism; learning from European cities. Island Press, Washington DC

Bettencourt LM (2013): The origin of scaling in cities. Science 340:1438–1441

Bibri E.S, Krogstie. J and Kanholm. M (2020): Compact City Planning and Development. Emerging Practices and Strategies for Achieving the goals of Sustainability.

Burak.G, Lwasa. S, Masundire.H, Parnell.S and Seto.K (2007): Urbanization in Africa: Challenges and Opportunities for Conservation.

Chenwel.M, Qirui. S and Yang. L (2023): One Community at a time. Promoting Community Resilience in the face of Natural Hazards and Public Health Challenge.

Daneshpour A, and Shakibamanesh A (2011): The compact city; does it create an obligatory context for urban sustainability? Int J Archit Eng Urban Plan 21(2):110–118

Dodoman.D, Leck.H, Rusca.M and Colen.S (2017): African Urbanization and Urbanism Implications for Risk Accumulation and Reduction. International Journal of Disaster Risk Reduction.

Donatiello JE (2015): The world's population: an encyclopedia of critical issues, crises, and ever-growing countries.

Drury F (2015): Enchanting photos of the abandoned village taken over by hippies living in fairytale houses in the Spanish mountains

Farmer P, Frojmovic M, Hague C, Harridge C, Narang S, Shishido R, Siegel D, Taylor P, Vogelij J (2006): Reinventing planning: a new governance paradigm for managing human settlements. Position paper for debate leading into the World Planners Congress, Vancouver

Farr D (2007): Sustainable urbanism: urban design with nature. Wiley, New York

Galal, S (2023): Urbanization Rate in Africa in 2021 by Country.

https://www.statista.com/statistics/1223543/urbanization-rate-in-africa-by-country

Galal. S (2023): Cost of living in selected cities as of 2023. https://www.statista.com/statistics/1218516/cost-of-living-in-selected-african-cities/

Girardet H (1996): The GAIA atlas of cities: new directions for sustainable urban living. Gaia Books Limited, London

Habitat III (2016): Revised zero draft of the new urban agenda. https://www.habitat3.org/zerodraft

Haferkamp.H and Smelser. N (1992): Social Change and Modernity. University of California Press

Howard E (1902): Garden cities of tomorrow. S. Sonnenschein & Co., Ltd, London

Kennedy G (2009): Food security in the context of urban sub-Saharan Africa. Internet paper for food security, food Africa, internet forum.

Konrad. A,S (2020): Addressing the Challenge of Urbanization in Africa. A Summary of the 2019 Policy Circle Discussions, Global Policy Institute.

McMichael AJ (2000): The urban environment and health in a world of increasing globalization: issues for developing countries. Bulletin of the World Health Organization.

Moroni S (2015) Complexity and the inherent limits of explanation and prediction: urban codes for self-organizing systems. Plan Theory 14(3):248–267

Moroni, S (2016): Urban Density after Jane Jacobs: The Crucial Role of Diversity and Emergence.

Msuya. I, Moshi.I and Levira.F (2020): Dar es Salaam: The Unplanned Urban Sprawl threatening neighborhood sustainability.

Neuman M (2005): The compact city fallacy. J Plan Educ Res 25:11–26. doi:10.1177/0739456X04270466

Okafor. C (2023): Top 5 African cities with the worst traffic. https://africa.businessinsider.com/local/markets/top-5-african-cities-with-the-worst-traffic/ytjzwyg

Ono. H and Tetsuo. K (2020): Understanding the development patterns of informal settlements in Nairobi, Kenya.

Ono.H and Kidokoro,T (2020):Understanding the Development Patterns of Informal Settlements in Nairobi

Orufa. S (2023): African Countries with the highest and lowest cost of living in 2023. https://venturesafrica.com/african-countries-with-the-highest-and-lowest-cost-of-living-in-2023/

Portugali J, Meyer H, Stolk E, Tan E (eds) (2012): Complexity theories of cities have come of age

Ramanatha, J (2019): Challenges of Urbanization in Africa.

Raptis, V.J (2017): Urbanism in the Making. A Handbook of Survival

Roggema R (2012): Swarm planning: the development of a spatial planning methodology to deal with climate adaptation, PhD thesis. Delft University of Technology, Delft

Roggema R (2014): Towards enhanced resilience in city design: a proposition. Land 3:460–481

Romero-Lankao, P; Dodman, D (2011): Cities in transition: Transforming urban centers from hotbeds of GHG emissions and vulnerability to seedbeds of sustainability and resilience. Introduction and Editorial overview. Curr. Opin. Environ. Sustain. 2011, 3, 113–120.

Soja, E.W.C (2009): The City and Spatial Justice

UNESCO (2015): An Indicative Review of UNESCO's Work on Social Inclusion: A UNESCO-ODI Study of Promoting Practices in the Field of Social Inclusion.

UN-Habitat (2020): Analysis of Multiple Deprivations in Secondary Cities in Sub-Saharan Africa.

United Nations (2018): Transforming Our World: The 2030 Agenda for Sustainable Development. United Nations Sustainable

United Nations, Department of Economic and Social Affairs, Population Division (2019): World Population Prospects 2019.

Vlatiov. D, Bounfford.J, Pearson.C and Morris.L (2010): Urban Health Global Perspectives

Wandera, P (2014): The political ecology of unplanned land use changes in Kampala city, Uganda.

TOWARDS SUSTAINABLE URBANISM

This book views transition to sustainable urbanism as a new urban agenda for rapidly urbanizing Africa. In a sense, the book kick-starts the process of determining pathways from urbanism to sustainable urbanism for Africa's rapidly urbanizing.

It is authored by **Dr. Daniel Wandera Clief Naku (PhD),** a lecturer in the Department of Geography at the St John's University of Tanzania, Dodoma. He holds a First Class Hons Degree in BSc Urban and Regional Planning from the University of Dar Es Salaam (UDSM), Tanzania; a Master of Science Degree in Development Planning and Management from TU-Dortmund, Germany and Kwame Nkrumah University of Science and Technology (KNUST); a Master Degree in Development Economics from the Institute of Rural Development Planning (IRDP), Tanzania as well as a PhD in Policy Analysis and Programme Management from Ardhi University (ARU), Tanzania.

He has over 15 years of working experience which involve university teaching, research, consulting and providing professional training. Besides this book, Dr. Wandera has also authored and published four books titled; *Project Analysis and Evaluation for Development Practitioners and Academicians: A Guide to Project Analysis and Evaluation of Development Projects; Going Strategic at the Community Level: An insight into the how's of strategic planning at the community level; Project Concept Note and Proposal Writing Guide: The Art of Writing Project Concept Notes and Proposals for Development Projects as well as Extending the Frontiers of Knowledge through Research and Publications. A Research Repository Book.*

FSC
www.fsc.org
MIX
Papier | Fördert
gute Waldnutzung
FSC® C083411

Kolibri 360 GmbH
Ferdinand-Jühlke-Straße 7
99095 Erfurt, Deutschland
produktsicherheit@kolibri360.de

Druck:
CPI Druckdienstleistungen GmbH
im Auftrag der
Kolibri 360 GmbH
Ferdinand-Jühlke-Str. 7
99095 Erfurt